Some numbers with many prime

210	496	27720	720	2320	1440	3600	576	1296	4096
8128	5040	1728	1984	1470	2808	900	1980	4851	1155
40320	960	1764	3080	6720	216	640	4480	2268	1870
7392	1050	2205	5733	5184	4368	1768	1694	4158	30030
8820	52003	17340	26026	1120	1225	1352	2210	4056	1734
2592	9464	2888	8748	4802	6936	1568	2048	19208	4212
14400	7735	14406	8232	8775	3888	3234	675	22542	3696
252	5488	2312	5832	8664	12636	2704	2744	16184	8670

Some numbers with factors only close to the square root

143	323	667	899	1763	3599	4087	5183	6557	247
221	391	551	713	851	1147	1073	2021	2773	3763
1591	187	4399	2747	1961	4189	3551	2173	2419	2279

Some fractions which cancel satisfyingly

$\frac{147}{343}$	$\frac{252}{294}$	$\frac{273}{357}$	$\frac{322}{406}$	$\frac{576}{792}$	$\frac{616}{1001}$	$\frac{714}{798}$	$\frac{7854}{11781}$
$\frac{153}{204}$	$\frac{105}{147}$	$\frac{182}{1092}$	$\frac{858}{273}$	$\frac{432}{512}$	$\frac{144}{360}$	$\frac{420}{525}$	$\frac{152}{361}$
$\frac{978}{652}$	$\frac{288}{432}$	$\frac{168}{408}$	$\frac{322}{483}$	$\frac{216}{288}$	$\frac{243}{729}$	$\frac{484}{847}$	$\frac{225}{275}$
$\frac{336}{924}$	$\frac{729}{1215}$	$\frac{392}{441}$	$\frac{256}{384}$	$\frac{245}{294}$	$\frac{343}{2401}$	$\frac{192}{576}$	$\frac{432}{468}$
$\frac{768}{2496}$	$\frac{128}{512}$	$\frac{315}{630}$	$\frac{504}{792}$	$\frac{252}{588}$	$\frac{336}{840}$	$\frac{252}{336}$	$\frac{588}{672}$

$\frac{500,094}{583,443}$ $\frac{132,132}{330,330}$ $\frac{94,962}{329,574}$

Squares of integers to 100

n	n²	n	n²	n	n²	n	n²
1	1	26	676	51	2601	76	5776
2	4	27	729	52	2704	77	5929
3	9	28	784	53	2809	78	6084
4	16	29	841	54	2916	79	6241
5	25	30	900	55	3025	80	6400
6	36	31	961	56	3136	81	6561
7	49	32	1024	57	3249	82	6724
8	64	33	1089	58	3364	83	6889
9	81	34	1156	59	3481	84	7056
10	100	35	1225	60	3600	85	7225
11	121	36	1296	61	3721	86	7396
12	144	37	1369	62	3844	87	7569
13	169	38	1444	63	3969	88	7744
14	196	39	1521	64	4096	89	7921
15	225	40	1600	65	4225	90	8100
16	256	41	1681	66	4356	91	8281
17	289	42	1764	67	4489	92	8464
18	324	43	1849	68	4624	93	8649
19	361	44	1936	69	4761	94	8836
20	400	45	2025	70	4900	95	9025
21	441	46	2116	71	5041	96	9216
22	484	47	2209	72	5184	97	9409
23	529	48	2304	73	5329	98	9604
24	576	49	2401	74	5476	99	9801
25	625	50	2500	75	5625	100	10000

Cubes of integers to 100

n	n³	n	n³	n	n³	n	n³
1	1	26	17576	51	132651	76	438976
2	8	27	19683	52	140608	77	456533
3	27	28	21952	53	148877	78	474552
4	64	29	24389	54	157464	79	493039
5	125	30	27000	55	166375	80	512000
6	216	31	29791	56	175616	81	531441
7	343	32	32768	57	185193	82	551368
8	512	33	35937	58	195112	83	571787
9	729	34	39304	59	205379	84	592704
10	1000	35	42875	60	216000	85	614125
11	1331	36	46656	61	226981	86	636056
12	1728	37	50653	62	238328	87	658503
13	2197	38	54872	63	250047	88	681472
14	2744	39	59319	64	262144	89	704969
15	3375	40	64000	65	274625	90	729000
16	4096	41	68921	66	287496	91	753571
17	4913	42	74088	67	300763	92	778688
18	5832	43	79507	68	314432	93	804357
19	6859	44	85184	69	328509	94	830584
20	8000	45	91125	70	343000	95	857375
21	9261	46	97336	71	357911	96	884736
22	10648	47	103823	72	373248	97	912673
23	12167	48	110592	73	389017	98	941192
24	13824	49	117649	74	405224	99	970299
25	15625	50	125000	75	421875	100	1000000

Fourth powers of integers to 100

n	n⁴	n	n⁴	n	n⁴	n	n⁴
1	1	26	456976	51	6765201	76	33362176
2	16	27	531441	52	7311616	77	35153041
3	81	28	614656	53	7890481	78	37015056
4	256	29	707281	54	8503056	79	38950081
5	625	30	810000	55	9150625	80	40960000
6	1296	31	923521	56	9834496	81	43046721
7	2401	32	1048576	57	10556001	82	45212176
8	4096	33	1185921	58	11316496	83	47458321
9	6561	34	1336336	59	12117361	84	49787136
10	10000	35	1500625	60	12960000	85	52200625
11	14641	36	1679616	61	13845841	86	54700816
12	20736	37	1874161	62	14776336	87	57289761
13	28561	38	2085136	63	15752961	88	59969536
14	38416	39	2313441	64	16777216	89	62742241
15	50625	40	2560000	65	17850625	90	65610000
16	65536	41	2825761	66	18974736	91	68574961
17	83521	42	3111696	67	20151121	92	71639296
18	104976	43	3418801	68	21381376	93	74805201
19	130321	44	3748096	69	22667121	94	78074896
20	160000	45	4100625	70	24010000	95	81450625
21	194481	46	4477456	71	25411681	96	84934656
22	234256	47	4879681	72	26873856	97	88529281
23	279841	48	5308416	73	28398241	98	92236816
24	331776	49	5764801	74	29986576	99	96059601
25	390625	50	6250000	75	31640625	100	100000000

Fifth powers of integers to 30

n	n⁵	n	n⁵	n	n⁵
1	1	11	161051	21	4084101
2	32	12	248832	22	5153632
3	243	13	371293	23	6436343
4	1024	14	537824	24	7962624
5	3125	15	759375	25	9765625
6	7776	16	1048576	26	11881376
7	16807	17	1419857	27	14348907
8	32768	18	1889568	28	17210368
9	59049	19	2476099	29	20511149
10	100000	20	3200000	30	24300000

Sixth powers of integers to 30

n	n⁶	n	n⁶	n	n⁶
1	1	11	1771561	21	85766121
2	64	12	2985984	22	113379904
3	729	13	4826809	23	148035889
4	4096	14	7529536	24	191102976
5	15625	15	11390625	25	244140625
6	46656	16	16777216	26	308915776
7	117649	17	24137569	27	387420489
8	262144	18	34012224	28	481890304
9	531441	19	47045881	29	594823321
10	1000000	20	64000000	30	729000000

The first 200 triangle numbers

1 3 6 10

The n´th triangle number is $\frac{n(n+1)}{2}$

n		n		n		n	
1	1	51	1326	101	5151	151	11476
2	3	52	1378	102	5253	152	11628
3	6	53	1431	103	5356	153	11781
4	10	54	1485	104	5460	154	11935
5	15	55	1540	105	5565	155	12090
6	21	56	1596	106	5671	156	12246
7	28	57	1653	107	5778	157	12403
8	36	58	1711	108	5886	158	12561
9	45	59	1770	109	5995	159	12720
10	55	60	1830	110	6105	160	12880
11	66	61	1891	111	6216	161	13041
12	78	62	1953	112	6328	162	13203
13	91	63	2016	113	6441	163	13366
14	105	64	2080	114	6555	164	13530
15	120	65	2145	115	6670	165	13695
16	136	66	2211	116	6786	166	13861
17	153	67	2278	117	6903	167	14028
18	171	68	2346	118	7021	168	14196
19	190	69	2415	119	7140	169	14365
20	210	70	2485	120	7260	170	14535
21	231	71	2556	121	7381	171	14706
22	253	72	2628	122	7503	172	14878
23	276	73	2701	123	7626	173	15051
24	300	74	2775	124	7750	174	15225
25	325	75	2850	125	7875	175	15400
26	351	76	2926	126	8001	176	15576
27	378	77	3003	127	8128	177	15753
28	406	78	3081	128	8256	178	15931
29	435	79	3160	129	8385	179	16110
30	465	80	3240	130	8515	180	16290
31	496	81	3321	131	8646	181	16471
32	528	82	3403	132	8778	182	16653
33	561	83	3486	133	8911	183	16836
34	595	84	3570	134	9045	184	17020
35	630	85	3655	135	9180	185	17205
36	666	86	3741	136	9316	186	17391
37	703	87	3828	137	9453	187	17578
38	741	88	3916	138	9591	188	17766
39	780	89	4005	139	9730	189	17955
40	820	90	4095	140	9870	190	18145
41	861	91	4186	141	10011	191	18336
42	903	92	4278	142	10153	192	18528
43	946	93	4371	143	10296	193	18721
44	990	94	4465	144	10440	194	18915
45	1035	95	4560	145	10585	195	19110
46	1081	96	4656	146	10731	196	19306
47	1128	97	4753	147	10878	197	19503
48	1176	98	4851	148	11026	198	19701
49	1225	99	4950	149	11175	199	19900
50	1275	100	5050	150	11325	200	20100

Pascal's triangle

1

1 1

1 2 1

1 3 3 1

1 4 6 4 1

1 5 10 10 5 1

1 6 15 20 15 6 1

1 7 21 35 35 21 7 1

1 8 28 56 70 56 28 8 1

1 9 36 84 126 126 84 36 9 1

1 10 45 120 210 252 210 120 45 10 1

1 11 55 165 330 462 462 330 165 55 11 1

1 12 66 220 495 792 924 792 495 220 66 12 1

Points to note

1. The numbers in each row are the coefficients of the expansion of $(a+b)^n$
2. Each number is the sum of the two numbers above.
3. The outer layer of numbers are all 1.
4. The next layer are consecutive integers.
5. The next layer are triangle numbers.
6. The sum of the numbers in row n is 2^n.
7. The sum of all the numbers above row n is 2^n-1.
8. All the numbers except for the 1 at each end of row n are divisible by n if n is prime.

A curious triangle

1 2 3 4 5 6 7 8

3 5 7 9 2 4 6

8 3 7 2 6 1

2 1 9 8 7

3 1 8 6

4 9 5

4 5

9

Start this triangle by writing the numbers 1 to 8 in order. Add each pair and write the sum in the space below. If the number is larger than 9 take the digital root by adding together the digits. For example 11 becomes 2, 13 becomes 4 and 15 becomes 6. Continue this process to the bottom. Note that there are 36 digits to complete the triangle and that each digit occurs four times exactly. It is very effective to perform this ´trick´ with a pack of cards from which all the tens and court cards have been removed. Then all the cards are used up as the triangle is completed.

Factorials

A factorial number is the product of the integer with all smaller integers down to 1. For instance 6! = 6x5x4x3x2x1. Note the use of the exclamation mark. 6! is read as 'six factorial'.
Common sense might say the 0! = 0, but it must in fact be 1.

Factorial numbers rapidly become very large as the value of n increases.

0!	1
1!	1
2!	2
3!	6
4!	24
5!	120
6!	720
7!	5,040
8!	40,320
9!	362,880
10!	3,628,800
11!	39,916,800
12!	479,001,600
13!	6,227,020,800
14!	87,178,291,200
15!	1,307,674,368,000
16!	20,922,789,888,000
17!	355,687,428,096,000
18!	6,402,373,705,728,000
19!	121,645,100,408,832,000

Prime numbers

A prime number is a number which has no factors other than itself and 1. It is not customary to include 1 itself in a list of primes and so the first prime number is 2. This first prime is also the only even prime and so we need only search odd numbers and only test with odd factors. Since it is true that any number which will divide by (say) 15 will also divide by 3 and 5, it is only necessary to test-divide by prime factors. Also, if prime factors are tested in ascending order, it is only necessary to go as far as the square root of the number being tested. Any prime factor larger than the square root would imply another prime factor smaller than the square root and this would have been found already.

The sieve of Eratosthenes

The Greek mathematician Eratosthenes in the third century BC showed that prime numbers can be found by ´sieving out´ numbers which have factors. One can start by crossing out all factors of 2, except for 2 itself, all factors of 3 except for 3 itself, all factors of 5 except for 5 and so on. The numbers which remain not crossed out are the primes.

	2	3	~~4~~	5	~~6~~	7	~~8~~	~~9~~	~~10~~
11	~~12~~	13	~~14~~	~~15~~	~~16~~	17	~~18~~	19	~~20~~
~~21~~	~~22~~	23	~~24~~	~~25~~	~~26~~	~~27~~	~~28~~	29	~~30~~
31	~~32~~	~~33~~	~~34~~	~~35~~	~~36~~	37	~~38~~	~~39~~	~~40~~
41	~~42~~	43	~~44~~	~~45~~	~~46~~	47	~~48~~	~~49~~	~~50~~
~~51~~	~~52~~	53	~~54~~	~~55~~	~~56~~	~~57~~	~~58~~	59	~~60~~
61	~~62~~	~~63~~	~~64~~	~~65~~	~~66~~	67	~~68~~	~~69~~	~~70~~
71	~~72~~	73	~~74~~	~~75~~	~~76~~	~~77~~	~~78~~	79	~~80~~
~~81~~	~~82~~	83	~~84~~	~~85~~	~~86~~	~~87~~	~~88~~	89	~~90~~
~~91~~	~~92~~	~~93~~	~~94~~	~~95~~	~~96~~	97	~~98~~	~~99~~	~~100~~

This sieve gives the primes to 100.

Twin primes

An examination of any list of prime numbers will soon identify "twins" or pairs of primes which differ by two.

Up to 100 there are eight such twins.

3 & 5 5 & 7 11 & 13 17 & 19 29 & 31 41 & 43 59 & 61 71 & 73

Between 100 and 200 there are seven.

101 & 103 107 & 109 137 & 139 149 & 151 179 & 181 191 & 193 197 & 199

Between 200 and 300 there are four

227 & 229 239 & 241 269 & 271 281 & 283

Between 300 and 500 there are five
311 & 313 347 & 349 419 & 421 431 & 433 461 & 463

Twin primes seem to occur throughout the number range and it is a simple matter to modify a program which searches for primes so that it identifies twins as **well.**

There are 8169 twin primes below a million.

Primes to 5,000

2	3	5	7	11	13	17	19	23	29	31	37
41	43	47	53	59	61	67	71	73	79	83	89
97	101	103	107	109	113	127	131	137	139	149	151
157	163	167	173	179	181	191	193	197	199	211	223
227	229	233	239	241	251	257	263	269	271	277	281
283	293	307	311	313	317	331	337	347	349	353	359
367	373	379	383	389	397	401	409	419	421	431	433
439	443	449	457	461	463	467	479	487	491	499	503
509	521	523	541	547	557	563	569	571	577	587	593
599	601	607	613	617	619	631	641	643	647	653	659
661	673	677	683	691	701	709	719	727	733	739	743
751	757	761	769	773	787	797	809	811	821	823	827
829	839	853	857	859	863	877	881	883	887	907	911
919	929	937	941	947	953	967	971	977	983	991	997

1009	1013	1019	1021	1031	1033	1039	1049	1051	1061	1063	1069
1087	1091	1093	1097	1103	1109	1117	1123	1129	1151	1153	1163
1171	1181	1187	1193	1201	1213	1217	1223	1229	1231	1237	1249
1259	1277	1279	1283	1289	1291	1297	1301	1303	1307	1319	1321
1327	1361	1367	1373	1381	1399	1409	1423	1427	1429	1433	1439
1447	1451	1453	1459	1471	1481	1483	1487	1489	1493	1499	1511
1523	1531	1543	1549	1553	1559	1567	1571	1579	1583	1597	1601
1607	1609	1613	1619	1621	1627	1637	1657	1663	1667	1669	1693
1697	1699	1709	1721	1723	1733	1741	1747	1753	1759	1777	1783
1787	1789	1801	1811	1823	1831	1847	1861	1867	1871	1873	1877
1879	1889	1901	1907	1913	1931	1933	1949	1951	1973	1979	1987
1993	1997	1999									

2003	2011	2017	2027	2029	2039	2053	2063	2069	2081	2083	2087
2089	2099	2111	2113	2129	2131	2137	2141	2143	2153	2161	2179
2203	2207	2213	2221	2237	2239	2243	2251	2267	2269	2273	2281
2287	2293	2297	2309	2311	2333	2339	2341	2347	2351	2357	2371
2377	2381	2383	2389	2393	2399	2411	2417	2423	2437	2441	2447
2459	2467	2473	2477	2503	2521	2531	2539	2543	2549	2551	2557
2579	2591	2593	2609	2617	2621	2633	2647	2657	2659	2663	2671
2677	2683	2687	2689	2693	2699	2707	2711	2713	2719	2729	2731
2741	2749	2753	2767	2777	2789	2791	2797	2801	2803	2819	2833
2837	2843	2851	2857	2861	2879	2887	2897	2903	2909	2917	2927
2939	2953	2957	2963	2969	2971	2999					

3001	3011	3019	3023	3037	3041	3049	3061	3067	3079	3083	3089
3109	3119	3121	3137	3163	3167	3169	3181	3187	3191	3203	3209
3217	3221	3229	3251	3253	3257	3259	3271	3299	3301	3307	3313
3319	3323	3329	3331	3343	3347	3359	3361	3371	3373	3389	3391
3407	3413	3433	3449	3457	3461	3463	3467	3469	3491	3499	3511
3517	3527	3529	3533	3539	3541	3547	3557	3559	3571	3581	3583
3593	3607	3613	3617	3623	3631	3637	3643	3659	3671	3673	3677
3691	3697	3701	3709	3719	3727	3733	3739	3761	3767	3769	3779
3793	3797	3803	3821	3823	3833	3847	3851	3853	3863	3877	3881
3889	3907	3911	3917	3919	3923	3929	3931	3943	3947	3967	3989

4001	4003	4007	4013	4019	4021	4027	4049	4051	4057	4073	4079
4091	4093	4099	4111	4127	4129	4133	4139	4153	4157	4159	4177
4201	4211	4217	4219	4229	4231	4241	4243	4253	4259	4261	4271
4273	4283	4289	4297	4327	4337	4339	4349	4357	4363	4373	4391
4397	4409	4421	4423	4441	4447	4451	4457	4463	4481	4483	4493
4507	4513	4517	4519	4523	4547	4549	4561	4567	4583	4591	4597
4603	4621	4637	4639	4643	4649	4651	4657	4663	4673	4679	4691
4703	4721	4723	4729	4733	4751	4759	4783	4787	4789	4793	4799
4801	4813	4817	4831	4861	4871	4877	4889	4903	4909	4919	4931
4933	4937	4943	4951	4957	4967	4969	4973	4987	4993	4999	

Primes 5,000 to 10,000

5003	5009	5011	5021	5023	5039	5051	5059	5077	5081	5087	5099
5101	5107	5113	5119	5147	5153	5167	5171	5179	5189	5197	5209
5227	5231	5233	5237	5261	5273	5279	5281	5297	5303	5309	5323
5333	5347	5351	5381	5387	5393	5399	5407	5413	5417	5419	5431
5437	5441	5443	5449	5471	5477	5479	5483	5501	5503	5507	5519
5521	5527	5531	5557	5563	5569	5573	5581	5591	5623	5639	5641
5647	5651	5653	5657	5659	5669	5683	5689	5693	5701	5711	5717
5737	5741	5743	5749	5779	5783	5791	5801	5807	5813	5821	5827
5839	5843	5849	5851	5857	5861	5867	5869	5879	5881	5897	5903
5923	5927	5939	5953	5981	5987						

6007	6011	6029	6037	6043	6047	6053	6067	6073	6079	6089	6091
6101	6113	6121	6131	6133	6143	6151	6163	6173	6197	6199	6203
6211	6217	6221	6229	6247	6257	6263	6269	6271	6277	6287	6299
6301	6311	6317	6323	6329	6337	6343	6353	6359	6361	6367	6373
6379	6389	6397	6421	6427	6449	6451	6469	6473	6481	6491	6521
6529	6547	6551	6553	6563	6569	6571	6577	6581	6599	6607	6619
6637	6653	6659	6661	6673	6679	6689	6691	6701	6703	6709	6719
6733	6737	6761	6763	6779	6781	6791	6793	6803	6823	6827	6829
6833	6841	6857	6863	6869	6871	6883	6899	6907	6911	6917	6947
6949	6959	6961	6967	6971	6977	6983	6991	6997			

7001	7013	7019	7027	7039	7043	7057	7069	7079	7103	7109	7121
7127	7129	7151	7159	7177	7187	7193	7207	7211	7213	7219	7229
7237	7243	7247	7253	7283	7297	7307	7309	7321	7331	7333	7349
7351	7369	7393	7411	7417	7433	7451	7457	7459	7477	7481	7487
7489	7499	7507	7517	7523	7529	7537	7541	7547	7549	7559	7561
7573	7577	7583	7589	7591	7603	7607	7621	7639	7643	7649	7669
7673	7681	7687	7691	7699	7703	7717	7723	7727	7741	7753	7757
7759	7789	7793	7817	7823	7829	7841	7853	7867	7873	7877	7879
7883	7901	7907	7919	7927	7933	7937	7949	7951	7963	7993	

8009	8011	8017	8039	8053	8059	8069	8081	8087	8089	8093	8101
8111	8117	8123	8147	8161	8167	8171	8179	8191	8209	8219	8221
8231	8233	8237	8243	8263	8269	8273	8287	8291	8293	8297	8311
8317	8329	8353	8363	8369	8377	8387	8389	8419	8423	8429	8431
8443	8447	8461	8467	8501	8513	8521	8527	8537	8539	8543	8563
8573	8581	8597	8599	8609	8623	8627	8629	8641	8647	8663	8669
8677	8681	8689	8693	8699	8707	8713	8719	8731	8737	8741	8747
8753	8761	8779	8783	8803	8807	8819	8821	8831	8837	8839	8849
8861	8863	8867	8887	8893	8923	8929	8933	8941	8951	8963	8969
8971	8999										

9001	9007	9011	9013	9029	9041	9043	9049	9059	9067	9091	9103
9109	9127	9133	9137	9151	9157	9161	9173	9181	9187	9199	9203
9209	9221	9227	9239	9241	9257	9277	9281	9283	9293	9311	9319
9323	9337	9341	9343	9349	9371	9377	9391	9397	9403	9413	9419
9421	9431	9433	9437	9439	9461	9463	9467	9473	9479	9491	9497
9511	9521	9533	9539	9547	9551	9587	9601	9613	9619	9623	9629
9631	9643	9649	9661	9677	9679	9689	9697	9719	9721	9733	9739
9743	9749	9767	9769	9781	9787	9791	9803	9811	9817	9829	9833
9839	9851	9857	9859	9871	9883	9887	9901	9907	9923	9929	9931
9941	9949	9967	9973								

Finding larger primes

For larger primes, the method of finding them remains the same but it needs a rather more powerful computer. Here are some primes which are a little larger than a trillion.

This table gives prime numbers of the form 1 000 000 000 000 + A, where A =

39	61	63	91	121	163	169	177	189	193
211	271	303	331	333	339	459	471	537	543
547	561	609	661	669	721	751	787	789	799
841	903	921	931	933	949	997			

There were 37 primes in this range, including three twins.

1 000 000 000 331 and 1 000 000 000 333
1 000 000 000 787 and 1 000 000 000 789
1 000 000 000 931 and 1 000 000 000 933

A very large prime

It is known that the number $2^{521} - 1$ is prime.

Written in the base 2, it has 521 digits

1111111111111111111111111111111111111111111111111111111111111111111111111111111
1111111111111111111111111111111111111111111111111111111111111111111111111111111
1111111111111111111111111111111111111111111111111111111111111111111111111111111
1111111111111111111111111111111111111111111111111111111111111111111111111111111
1111111111111111111111111111111111111111111111111111111111111111111111111111111
1111111111111111111111111111111111111111111111111111111111111111111111111111111
11111111111111111111111111111111111111111111111

Written in base 10 it has 157 digits

6864797660130609714981900799081393217269435300143305409394463459185543183397656
052122559640661454554977296311391480858037121987999716643812574028291115057151

A number of this size is almost beyond comprehension

It may be written as approximately 6.9×10^{157}

The number of human beings alive in the world today is approximately 6×10^{9}

The number of human beings there have ever been is of the order of 6×10^{10}

The age of the Universe is of the order 15×10^{9} years or 4.5×10^{17} seconds.

This prime number is in fact only the 13th of a series called Mersenne Primes and is minute by comparison with the ones which follow.

Mersenne primes

Father Mersenne was a French monk who in the 17th century studied numbers of the form

$$M_k = 2^k - 1$$

The first ten of these numbers yields 1,seven primes and two composite numbers.

k = 1	1	
k = 2	3	
k = 3	7	
k = 5	31	
k = 7	127	
k =11	2047	(which divides by 23)
k =13	8191	
k =17	131071	
k =19	524287	
k =23	8388607	(which divides by 47)

Now it is proved by Fermat and Euler that all factors of a Mersenne number must be simultaneously of the form 8n + 1 and 2kp - 1 where k and n are integers and p is prime.

$$M_{11} = 2^{11} - 1 = 2047 = 23 \times 89$$
where $23 = 2 \times 11 + 1$ and $89 = 8 \times 11 + 1$

Father Mersenne prophesied that the values of k

2,3,5,7,13,17,19,31,67,127,257

would all yield primes and that all other values of k less than 257 would yield composite numbers. However, he had no way of working out the larger ones or of proving his theorem. The arrival of the computer has shown that he was wrong about 67 and 257, and that 61, 89 and 107 do yield primes.

Primes of this form are called 'Mersenne Primes' and there is (relatively speaking!) a simple way of proving whether or not a Mersenne number is prime. It is therefore on Mersenne Primes that the search for the largest known prime has tended to concentrate. In 1979 the largest known prime was the Mersenne Prime with k = 44,497 which is a number with 13,395 digits. In 1982 David Slowinski showed that k = 86,243 yields a prime number with 25,962 digits and in 1983 that k = 132,049 yields a prime number with 39,751 digits. Since then a further eight have been discovered. The best way to find the current champion is to search the internet. Many things of interest will be discovered on the way.

The values of k for the 38 Mersenne Primes discovered so far are given in this table.

1	2	8	31	15	1279	22	9941	29	110503	36	2976221
2	3	9	61	16	2203	23	11213	30	132049	37	3021377
3	5	10	89	17	2281	24	19937	31	216091	38 (?)	6972593
4	7	11	107	18	3217	25	21701	32	756839		
5	13	12	127	19	4253	26	23209	33	859433		
6	17	13	521	20	4423	27	44497	34	1257787		
7	19	14	607	21	9689	28	86243	35	1398269		

At the time of writing the largest known prime is generated when k = 6972593. However, there may be undiscovered prime numbers in the interval between it and Mersenne Prime 37. See also page 15.

Modular arithmetic

If n and a are positive integers and if n divides into a with remainder r then we say

$$a \bmod n = r \qquad \text{so } 0 \leqslant a \bmod n < n$$

numerical examples

$17 \bmod 4 = 1 \qquad 28 \bmod 5 = 3 \qquad 30 \bmod 6 = 0 \qquad 102 \bmod 10 = 2$

Modular arithmetic obeys the following rules of addition subtraction and multiplication.

$$a \bmod c + b \bmod c = (a + b) \bmod c$$

$$a \bmod c - b \bmod c = (a - b) \bmod c$$

$$(a \bmod c) \times (b \bmod c) \bmod c = ab \bmod c$$

1. If $a_1 \bmod n = a_2 \bmod n = r$, then $(a_1 - a_2) \bmod n = 0$
   This means that if a_1 and a_2 give the same remainder when divided by n, then $a_1 - a_2$ has n as a factor.

2. If $a_1 \bmod n = r_1$ and $a_2 \bmod n = r_2$
   then $a_1 a_2 \bmod n = r_1 r_2 \bmod n$
   and $(a_1 a_2 - r_1 r_2) \bmod n = 0$

   Hence we can conclude that if $a_1, a_2, a_3, \ldots, a_n$ gives remainders $r_1, r_2, r_3, \ldots, r_n$ when divided by n, then $(a_1 a_2 a_3 \ldots a_n - r_1 r_2 r_3 \ldots r_n)$ divides by n.

Fermat's little theorem

The theorem states that for any integers $a \geqslant 2$
if n is prime then $(a^n - a) \bmod n = 0$

We exclude the case where a is divisible by n.

Let us consider the numbers

$$a, 2a, 3a, \ldots\ldots (n-1)a$$

If n is prime, it cannot divide into any of these numbers. The remainders obtained by these (n-1) divisions must all be different, because if two remainders were the same, the difference of those two numbers would divide by n. But the difference of those two numbers would be another term in the sequence and we know that none of them can divide by n if n is prime.
Hence the remainders must be 1,2,3,...,(n-1) in some order.

The number $a^{n-1}(n-1)! - (n-1)!$ is obtained by multiplying the numbers together and subtracting the product of the remainders. From the theorem of modular arithmetic shown above, this expression divides by n.
If n is prime, then (n-1)! does not divide by n. Hence $(a^{n-1}-1)$ does, and so does $a^n - a$

Hence $(a^n - a) \bmod n = 0$ if n is prime.

Pseudoprimes

Fermat´s little theorem states that if n is a prime number, then

$$(a^n - a) \bmod n = 0$$

It does not state that if the remainder is zero that n is necessarily prime. There is a small probability that the remainder can be zero, even though n is not a prime. Such numbers are called pseudoprimes.

Certain numbers, e.g. 15 and 341 pass the test for certain values of a.

The simplest example is for n = 15, a = 4

$4^{15} - 4 = 1,073,741,820$ which divides by 15 to give 71,582,788

The least value of n for which a = 2 is 341

$2^{341} - 2$ divides by 341 although 341 = 11 x 31

The existence of this value is the basis for an elegant proof that there is an infinite number of pseudoprimes in base 2.

Some other numbers pass the test for all values of a. Such numbers are called Carmichael numbers. Each consists of at least three prime factors. The sequence starts 561, 1105, 1729, 2465....

For checking primes on a computer, it is usual to put a = 2. In this case there are about 20,000 pseudoprimes out of nearly 900 million primes in the range up to 20 billion. This gives a probability that any number in this range which passes the test not being a prime is about one in 45,000.

Testing a very large number

We wish to test if R is prime where

R = 984,073,151,544,289,893,895,597.

Using Fermat´s little theorem for a = 2 we must calculate

$$(2^R - 2) \bmod R$$

If the remainder is not zero, then the number is certainly composite. If the remainder is zero, then there is a small probability that it is not prime.

A computer program was written to calculate $(2^R - 2) \bmod R$ and it found the remainder to be:

707,843,344,947,948,903,498,451

Hence we can be certain that R is not prime.

This method does not offer any indication what the factors are. In fact it was constructed by calculating the product of two prime numbers
984,073,142,047 and 1,000,000,009,651.

The study of very large prime numbers has been until now of largely academic interest. However, a new method of cryptography, called public key cryptography has been proposed which makes use of the product of two very large (more than 200 digits each) primes. No computer which exists or at present could be proposed could hope to factor such a large number in less than hundreds of years.

Counting the number of primes

Ever since prime numbers were first investigated people have been fascinated by trying to find a pattern in their occurrence. One has never been found, nor is it likely that one exists.

The sieve of Eratosthenes offers a very speedy method of finding primes and so a computer can rapidly count how many primes there are in any given range. The number of primes found in each block of 10,000 up to 1,000,000 is given in the table below. From this table it can be seen that about 12% of the first 10,000 numbers and about 10% of the next 10,000 numbers are prime.

1229	1033	983	958	930	924	878	902	876	879
861	848	858	851	838	835	814	845	828	814
823	811	819	784	823	793	805	790	792	773
803	808	796	778	795	780	765	778	767	793
754	776	772	779	765	752	765	782	761	772
753	770	764	747	750	750	747	769	763	747
763	751	729	733	757	733	745	754	752	728
763	723	760	742	707	740	755	735	738	745
732	733	745	729	727	725	753	728	732	719
752	708	740	713	720	711	732	717	710	721

Total number of primes 78498

Notice that there is a steady overall decline in the number of primes found in each block of 10,000, but that the number fluctuates in a seemingly random way.

It has been shown that the probability that a number x is prime is approximately $1/\log_e x$

Hence for any given range the number of primes may be estimated approximately. For the range 2 to 1,000,000, the approximate number is estimated to be about 78628. This compares well with the exact number of 78,498 obtained by listing and counting them all.

Goldbach's theorem

Goldbach stated that any even number may be expressed as the sum of at most two primes. It has not so far been proved, although no example has ever been found of a number which disproves it.

In a teaching situation it is interesting to propose the (false) conjecture that ´any number is either prime or is the sum of two primes´. Ignoring 1, it works for 90 out of the first 100 numbers.

Counter examples include

27, 35, 51, 57, 65, 77, 87, 93, 95

Perfect numbers

A "perfect" number is one which is equal to the sum of its divisors including one. Probably the idea of "perfection" is based on mysticism and the Pythagorean idea of the significance of certain numbers.

The first four perfect numbers are

1. 6 = 1+2+3
2. 28 = 1+2+4+7+14
3. 496 = 1+2+4+8+16+31+62+124+248
4. 8128 = 1+2+4+8+16+32+64+127+254+508+1016+2032+4064

Notice that the factors start off as ascending powers of 2,then there is a prime number,then that prime number is multiplied by successive powers of 2.

Since $6 = 2^1(2^2-1)$; $28 = 2^2(2^3-1)$; $496 = 2^4(2^5-1)$; $8128 = 2^6(2^7-1)$,this suggests that there is a method of identifying perfect numbers without a long and tedious search.

<u>Theorem</u> If 2^n-1 is prime, then $2^{n-1}(2^n-1)$ is perfect.

<u>Proof</u> Consider the divisors of $2^{n-1}(2^n-1)$

$$1,\ 2,\ 2^2,\ \ldots\ 2^{n-1} \text{ and } (2^n-1),\ 2(2^n-1),\ \ldots\ 2^{n-2}(2^n-1)$$

There are no other divisors since 2^n-1 is prime.

Summing the divisors

$$1+2+2^2\ldots2^{n-1} = 2^n-1$$

$$(2^n-1)(1+2+2^2\ldots2^{n-2}) = (2^{n-1})(2^n-1)-(2^n-1)$$

$$\text{TOTAl} = 2^{n-1}(2^n-1)$$

It has been proved that there are no other even perfect numbers.

There could be odd perfect numbers,but searches have shown that there are none as far as 10^{100}

The next four perfect numbers are

5 33550336 (when n = 13) 7 137468691328 (when n = 19)

6 8589689056 (when n = 17) 8 2305843008139952128 (when n = 31)

Some 37 more perfect numbers are known at present. The values of n are

9	61	14	607	19	4253	24	19937	28	86243	33	859433
10	89	15	1279	20	4423	25	21701	29	110503	34	1257787
11	107	16	2203	21	9689	26	23209	30	132049	35	1398269
12	127	17	2281	22	9941	27	44497	31	216091	36	2976221
13	521	18	3217	23	11213	28	86243	32	756839	37	3021377

A further value when n = 6972593 has been discovered but it is not clear yet whether this is number 38 or if there is an undiscovered perfect number in the interval. Indeed, there could be more than one. Further discoveries await the finding of yet larger prime numbers of the form $2^n - 1$.

Pythagorean triples

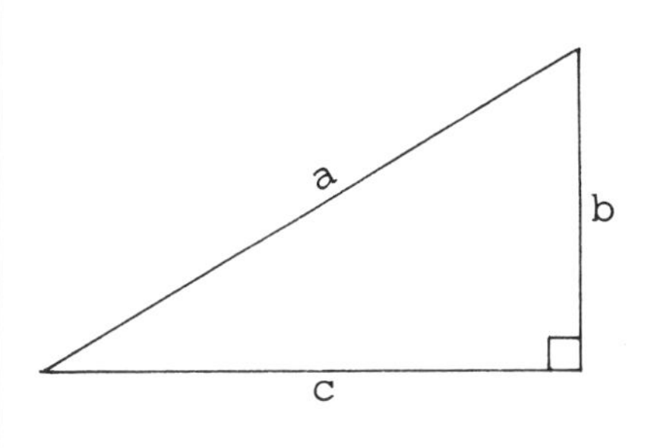

The theorem of Pythagoras states that in a right-angled triangle the square on the hypotenuse is equal to the sum of the squares on the other two sides.

$$a^2 = b^2 + c^2$$

Certain right angled triangles have all three sides as integers and these are of particular interest. The best known example is 3,4,5 but other well known ones are 5,12,13 and 6,8,10. However the triangle 6,8,10 is a multiple of 3,4,5 and is therefore the same shape and cannot be considered distinct.

This list contains all 158 distinct right-angled triangles, not counting multiples, where all three sides are less than 1000.

5	4	3	265	264	23	505	456	217	761	760	39
13	12	5	265	247	96	505	377	336	769	600	481
17	15	8	269	260	69	509	459	220	773	748	195
25	24	7	277	252	115	521	440	279	785	783	56
29	21	20	281	231	160	533	525	92	785	736	273
37	35	12	289	240	161	533	435	308	793	775	168
41	40	9	293	285	68	541	420	341	793	665	432
53	45	28	305	273	136	545	544	33	797	572	555
61	60	11	305	224	207	545	513	184	809	759	280
65	63	16	313	312	25	557	532	165	821	700	429
65	56	33	317	308	75	565	493	276	829	629	540
73	55	48	325	323	36	565	403	396	841	840	41
85	84	13	325	253	204	569	520	231	845	837	116
85	77	36	337	288	175	577	575	48	845	836	123
89	80	39	349	299	180	593	465	368	853	828	205
97	72	65	353	272	225	601	551	240	857	825	232
101	99	20	365	364	27	613	612	35	865	816	287
109	91	60	365	357	76	617	608	105	865	703	504
113	112	15	373	275	252	625	527	336	877	805	348
125	117	44	377	352	135	629	621	100	881	800	369
137	105	88	377	345	152	629	460	429	901	899	60
145	144	17	389	340	189	641	609	200	901	780	451
145	143	24	397	325	228	653	572	315	905	777	464
149	140	51	401	399	40	661	589	300	905	663	616
157	132	85	409	391	120	673	552	385	925	924	43
169	120	119	421	420	29	677	675	52	925	756	533
173	165	52	425	416	87	685	684	37	929	920	129
181	180	19	425	304	297	685	667	156	937	912	215
185	176	57	433	408	145	689	680	111	941	741	580
185	153	104	445	437	84	689	561	400	949	900	301
193	168	95	445	396	203	697	672	185	949	851	420
197	195	28	449	351	280	697	528	455	953	728	615
205	187	84	457	425	168	701	651	260	965	957	124
205	156	133	461	380	261	709	660	259	965	884	387
221	220	21	481	480	31	725	644	333	977	945	248
221	171	140	481	360	319	725	627	364	985	864	473
229	221	60	485	483	44	733	725	108	985	697	696
233	208	105	485	476	93	745	713	216	997	925	372
241	209	120	493	475	132	745	624	407			
257	255	32	493	468	155	757	595	468			

A hypotenuse in two ways

A search through the list shows that there are many examples where the hypotenuse can be obtained in two different ways.

The first pair is	65	63	16
	65	56	33
the second is	85	84	13
	85	77	36
and the third is	145	144	17
	145	143	24

It is easy to prove that any number which can be expressed as the product of the sums of squares can itself be expressed as the sums of squares in two different ways.

$$(p^2+q^2)\ (r^2+s^2)=p^2r^2+q^2r^2+p^2s^2+q^2s^2$$
$$= (pr-qs)^2 + (ps+qr)^2$$
$$= (pr+qs)^2 + (ps-qr)^2$$

The number 65 may be expressed as 5 x 13 where 5 = 1+4 and 13 = 4+9

The number 85 may be expressed as 5 x 17 where 5 = 1+4 and 17 = 1+16

The number 145 may be expressed as 5 x 29 where 5 = 1+4 and 29 = 4+25

A hypotenuse in four ways

It is not difficult to prove that numbers which are themselves of the form

$$(p^2+q^2)(r^2+s^2)(t^2+u^2)$$

may be expressed as the sums of squares in four different ways.

There are no such numbers under 1000
The smallest example is 1105, which is 5x13x17
where 5 = 1+4, 13 = 4+9, 17 = 1+16

The four triangles are

1105	1104	47
1105	1073	264
1105	943	576
1105	817	744

There are another three triples which have hypotenuses of 1105, but they are multiples. They are:

1105	884	663	(multiples of 221)
1105	1020	425	(multiples of 85)
1105	975	520	(multiples of 65)

Generating Pythagorean triples

While the first Pythagorean triples were undoubtedly found by trial and error, it is much more interesting to devise a method which will generate them.

If $a = p^2 + q^2$ $\quad b = p^2 - q^2$ $\quad c = 2pq$

Then clearly if p,q are integers and if p>q, then a,b,c are positive integers and $a^2=b^2+c^2$. We could therefore use different combinations of p,q to generate Pythagorean triples.

For example, if	p=2	q=1	a= 5,	b= 3,	c= 4
	p=3	q=1	a=10,	b= 8,	c= 6
	p=3	q=2	a=13,	b= 5,	c=12

The second triangle is simply a multiple of the first and will be the same shape. It would be preferable to eliminate such multiples, ideally by divising some simple rule about which values of p and q to try.

1. If we ignore multiples, then a must always be odd. We can argue that if a right-angled triangle exists where a is even and b,c are integers, then a triangle half the size would have sides a/2, b/2, c/2. Then a/2 would be an integer, but if multiples were excluded then both b/2 and c/2 would, because b,c must both be odd, both end in a half. The sums of the squares of these sides could not equal the square on the hypotenuse. Hence no such triangle exists, and the hypotenuse must be odd. If a is odd, then one of p,q must be even and the other odd.
2. If p and q have a common factor, say x, then a,b,c will each have a common factor x^2. Since multiples are excluded then a,b,c do not have a common factor and so p and q must not have any common factors either.

Hence if we systematically work through integer values of p and q where p > q, choosing one to be even and one to be odd and ignoring pairs which share a common factor, then we can generate Pythagorean triples at will.

For example,if	p=2	q=1	a= 5,	b= 3,	c= 4
	p=3	q=2	a=13,	b= 5,	c=12
	p=4	q=1	a=17,	b=15,	c= 8
	p=4	q=3	a=25,	b= 7,	c=24

The question still remains whether there exist triangles with integer values of a,b,c where p,q are not integers.

Clearly $p^2 = (a + b)/2$ $\quad q^2 = (a - b)/2$

Now p and q are both integers if (a + b)/2 and (a - b)/2 are both perfect squares. Since $(a + b)(a - b)/4 = p^2 q^2 = (c/2)^2$, we can see that if one is not a perfect square , the other cannot be. Hence we need only to examine the case that both are not perfect squares.If neither is a perfect square and yet their product is a perfect square, then it follows that they must share a common factor (say x). Hence c/2 shares the same common factor x. Hence a,b,c all have the common factor x and the triangle is a multiple. But this possibility has been excluded and so we can reject the possibility that (a + b)/2 and (a - b)/2 are not perfect squares.Hence if a,b,c are integers with no common factor, then p and q will both be integers and the generating method will find all triples.

A curious proportion

THE NUMBER OF DISTINCT INTEGER RIGHT-ANGLED TRIANGLES WITH HYPOTENUSES LESS THAN N.

N	NUMBER	PROPORTION
10	1	0.1
100	16	0.16
1000	158	0.158
10000	1593	0.1593
100000	15919	0.15919
1000000	159139	0.159139

This investigation suggested that the proportion tends to a limit. Subsequent theoretical work showed that the number of distinct integer right-angled triangles with hypotenuse less than N tends to $N/2\pi$

$$1/2\pi = 0.159159...$$

Fermat's last theorem

After finding so many solutions to the equation $x^2 + y^2 = z^2$, it became natural to look for whole number solutions to $x^3 + y^3 = z^3$. Rather to the surprise of everybody after a lot of searching, it proved impossible to find even one solution. The triple $x = 5$, $y = 6$ and $z = 7$ came fairly close but $125 + 216 = 241$ and not the 243 which is 7^3.

Then it was discovered that the equation $x^4 + y^4 = z^4$, did not seem to have any whole number solutions either. Nor did $x^5 + y^5 = z^5$ or $x^6 + y^6 = z^6$.

In about 1637, Pierre de Fermat proposed the conjecture that $x^n + y^n = z^n$ has no whole number solutions when n is greater than two.

In his notebook at the time, Fermat wrote: "I have discovered a truly marvellous proof which however the margin is not large enough to contain". This proof, if it existed, would have turned the conjecture into a theorem but it never came to light. What he thought of it himself we do not know as in the remaining eight years of his life he never referred to this theorem again.

After his death in 1641, his son published his theorems and conjectures and by the early eighteen hundreds all of them had either been proved or disproved with the exception of this one. That is why it became known as Fermat's last theorem and the search for a proof began. This search was to last for more than 350 years. Fermat himself had produced an acceptable proof for $n = 4$ and a development of his method was used to prove it for $n = 3$. In 1828 and 1830 two independent proofs for $n = 5$ were discovered and in 1842 one for $n = 7$. Two hundred years had now passed and valuable prizes were offered for an acceptable proof. Thousands of suggestions were sent in but all were shown to be faulty. The invention of the computer speeded up the process of testing for possible solutions. Just one set of three numbers which was true for any value of n would be sufficient to disprove it, but no example was ever found.

It was not until 1993 that Andrew Wiles presented a set of three lectures in Cambridge at which he proposed a proof that ran to more than 200 pages. Detailed examination by other mathematicians discovered a flaw and it required a further two years to deal with it. Then it could finally be said that Fermat's last theorem was indeed true. This proof was a real product of the twentieth century and was built upon the theorems and discoveries of many, many others. Certainly Fermat could not have done it this way. The balance of probability is that he was mistaken when he thought that he had proved it in 1637.

Fractions into decimals

FRACTION		DECIMAL TO 50 PLACES	RECURRING DIGITS
1/1	=	1	0
1/2	=	0.5	0
1/3	=	0.33333333333333333333333333333333333333333333333333	1
1/4	=	0.25	0
1/5	=	0.2	0
1/6	=	0.16666666666666666666666666666666666666666666666666	1
1/7	=	0.14285714285714285714285714285714285714285714285714	6*
1/8	=	0.125	0
1/9	=	0.11111111111111111111111111111111111111111111111111	1
1/10	=	0.1	0
1/11	=	0.09090909090909090909090909090909090909090909090909	2
1/12	=	0.08333333333333333333333333333333333333333333333333	1
1/13	=	0.07692307692307692307692307692307692307692307692307	6
1/14	=	0.07142857142857142857142857142857142857142857142857	6
1/15	=	0.06666666666666666666666666666666666666666666666666	1
1/16	=	0.0625	0
1/17	=	0.05882352941176470588235294117647058823529411764705	16*
1/18	=	0.05555555555555555555555555555555555555555555555555	1
1/19	=	0.05263157894736842105263157894736842105263157894736	18*
1/20	=	0.05	0
1/21	=	0.04761904761904761904761904761904761904761904761904	6
1/22	=	0.04545454545454545454545454545454545454545454545454	2
1/23	=	0.04347826086956521739130434782608695652173913043478	22*
1/24	=	0.04166666666666666666666666666666666666666666666666	1
1/25	=	0.04	0
1/26	=	0.03846153846153846153846153846153846153846153846153	6
1/27	=	0.03703703703703703703703703703703703703703703703703	3
1/28	=	0.03571428571428571428571428571428571428571428571428	6
1/29	=	0.03448275862068965517241379310344827586206896551724	28*
1/30	=	0.03333333333333333333333333333333333333333333333333	1
1/31	=	0.03225806451612903225806451612903225806451612903225	15
1/32	=	0.03125	0
1/33	=	0.03030303030303030303030303030303030303030303030303	2
1/34	=	0.02941176470588235294117647058823529411764705882352	16
1/35	=	0.02857142857142857142857142857142857142857142857142	6
1/36	=	0.02777777777777777777777777777777777777777777777777	1
1/37	=	0.02702702702702702702702702702702702702702702702702	3
1/38	=	0.02631578947368421052631578947368421052631578947368	18
1/39	=	0.02564102564102564102564102564102564102564102564102	6
1/40	=	0.025	0
1/41	=	0.02439024390243902439024390243902439024390243902439	5
1/42	=	0.02380952380952380952380952380952380952380952380952	6
1/43	=	0.02325581395348837209302325581395348837209302325581	21
1/44	=	0.02272727272727272727272727272727272727272727272727	2
1/45	=	0.02222222222222222222222222222222222222222222222222	1
1/46	=	0.02173913043478260869565217391304347826086956521739	22
1/47	=	0.02127659574468085106382978723404255319148936170212	46*
1/48	=	0.02083333333333333333333333333333333333333333333333	1
1/49	=	0.02040816326530612244897959183673469387755102040816	42
1/50	=	0.02	0

Cyclic decimals

To change a fraction into a decimal means dividing the numerator by the denominator. Certain fractions become decimals which recur. The best known example of this is 1/7

```
       .142857
 7) 1.0000000000
      7
      30
      28
       20
       14
        60
        56
         40
         35
          50
          49
           1
```

The remainders follow the sequence 3,2,6,4,5,1.Since a denominator 7 can only have a maximum of six different remainders, then the pattern must begin to ´cycle´ or repeat once they have all occurred . The number 7 belongs to a special class of numbers where all possible remainders do occur before the decimal does begin to cycle. Most recur in fewer than the maximum, as can be seen on the page opposite.

The ones where all possible remainders occur are marked with a star.

Numbers 1/n which repeat in (n-1) digits

There are 60 denominators less than 1000 which have this property.

7	17	19	23	29	47	59	61	97	109
113	131	149	167	179	181	193	223	229	233
257	263	269	313	337	367	379	383	389	419
433	461	487	491	499	503	509	541	571	577
593	619	647	659	701	709	727	743	811	821
823	857	863	887	937	941	953	971	977	983

Some curious relationships

$11^2 = 121$ and $11^3 = 1331$ and $11^4 = 14641$

$12^2 = 144$ and $21^2 = 441$; $13^2 = 169$ and $31^2 = 961$

$3^3 + 4^3 + 5^3 = 6^3$

$4^4 + 6^4 + 8^4 + 9^4 + 14^4 = 15^4$

$4^5 + 5^5 + 6^5 + 7^5 + 9^5 + 11^5 = 12^5$

Some curious approximations

π is approximately $\frac{22}{7}$, $\frac{355}{113}$, $\frac{104,348}{33,215}$, $\frac{312,689}{99,532}$

e is approximately $\frac{87}{32}$, $\frac{193}{71}$, $\frac{1,457}{536}$, $\frac{23,255}{8,544}$

$\sqrt{2}$ is approximately $\frac{99}{70}$, $\frac{239}{169}$, $\frac{577}{408}$, $\frac{19,601}{13,860}$

ϕ is approximately $\frac{89}{55}$, $\frac{610}{377}$, $\frac{4,181}{2,584}$, $\frac{75,025}{46,368}$

ϕ (phi) is the symbol usually used for the golden ratio

Some curious products

$$\prod_{p=2}^{\infty} \frac{p^2+1}{p^2-1} = \frac{5}{2}$$ where p is prime

$$\prod_{p=2}^{\infty} \left(1 - \frac{1}{p^2}\right) = \frac{6}{\pi^2}$$ where p is prime

$$\prod_{n=1}^{\infty} \left(1 - \frac{1}{(2n+1)^2}\right) = \frac{\pi}{4}$$ where n is an integer

It is interesting to program a computer to calculate each of these products successively and to watch each converge to its limit.

Fibonacci numbers

This sequence of numbers takes its name from a 13th century Italian mathematician who wrote under the name of Fibonacci. The original sequence started with 1,1,2,3,5,8, and was the solution to a problem about the total number of rabbits in an enclosure. The name is sometimes used to refer to any sequence where each number is the sum of the previous two. Sometimes sequences starting with numbers other than 1, 1 are called Lucas numbers. The first two numbers have to be stated and then the sequence is uniquely defined and can be carried on for ever. If the first two numbers are P and Q, then the sequence may be written

P, Q, P+2Q, 2P+3Q, 3P+5Q...

The ratio of consecutive terms rapidly converges to the golden ratio which is 1.61803398874989.....

P = 1, Q = 1		P = 2, Q = 7	
1		2	
1	1	7	3.5
2	2	9	1.285714285714
3	1.5	16	1.777777777777
5	1.666666666666	25	1.5625
8	1.6	41	1.64
13	1.625	66	1.609756097560
21	1.615384615384	107	1.621212121212
34	1.619047619047	173	1.616822429906
55	1.617647058823	280	1.618497109826
89	1.618181818181	453	1.617857142857
144	1.617977528089	733	1.618101545253
233	1.618055555555	1186	1.618008185538
377	1.618025751072	1919	1.618043844856
610	1.618037135278	3105	1.618030224075
987	1.618032786885	5024	1.618035426731
1597	1.618034447821	8129	1.618033439490
2584	1.618033813400	13153	1.618034198548
4181	1.618034055727	21282	1.618033908614
6765	1.618033963166	34435	1.618034019359
10946	1.618033998521	55717	1.618033977058
17711	1.618033985017	90152	1.618033993215
28657	1.618033990175	145869	1.618033987044
46368	1.618033988205	236021	1.618033989401
75025	1.618033988957	381890	1.618033988501
121393	1.618033988670	617911	1.618033988844
196418	1.618033988780	999801	1.618033988713
317811	1.618033988738	1617712	1.618033988763
514229	1.618033988754	2617513	1.618033988744

It is easy to program a computer to produce Fibonacci numbers for different values of P & Q and to calculate the ratio of consecutive terms.

The ratio converges to the golden ratio for all values of P, Q with the exception of P=1, Q = - 0.618033988....Try the effect of different numbers of decimal places of this number.

3 x 3 magic squares

A 3 x 3 magic square has 9 cells and so it is satisfying to find squares with the digits 1 to 9, each occurring once only.

There is only one such square possible and this has been known for a very long time. In China it is called the ´lo-shu´.

2	7	6
9	5	1
4	3	8

The sum of each row and column and each diagonal is 15

Of course it can be rotated and reflected to give magic squares which are apparently different, but really the same.

4	3	8
9	5	1
2	7	6

or

8	1	6
3	5	7
4	9	2

A unlimited number of 3 x 3 magic squares may be created by adding a constant to each cell.

Add 1 to give a total of 18.

5	4	9
10	6	2
3	8	7

or

Add 5 to give a total of 30

13	6	11
8	10	12
9	14	7

Of course there are many 3 x 3 magic squares which do not use consecutive numbers. Here are two which only use prime numbers. The first one classifies 1 as a prime number, the second does not.

67	1	43
13	37	61
31	73	7

101	29	83
53	71	89
59	113	41

A 3 x 3 multiplicative magic square

Although magic squares are normally formed by addition, it is possible to form them by multiplication. This 3 x 3 square gives the smallest possible product 216.

12	1	18
9	6	4
2	36	3

4 x 4 magic squares

A 4 x 4 magic square has 16 cells and so it is satisfying to find squares with the digits 1 to 16, each occurring once only.

There are 880 possible squares, not counting reflections and rotations and they have all been listed and classified by computer.

The simplest is developed from the set of numbers 1 to 16 placed in four rows.

1	2	3	4
5	6	7	8
9	10	11	12
13	14	15	16

This is not a magic square, but with just a few interchanges gives

1	15	14	4
12	6	7	9
8	10	11	5
13	3	2	16

The sums are all 34

There is a special class of 4 x 4 magic squares where not only do all the rows columms and diagonals add up to 34, but also so do each group of four numbers.These are known as DIABOLIC or PANDIAGONAL MAGIC SQUARES. There are 48 in all and they are listed below. Each may be rotated and reflected in 8 different ways.

1	8	10	15	1	8	10	15	1	8	11	14	1	8	11	14	1	8	13	12	1	8	13	12
12	13	3	6	14	11	5	4	12	13	2	7	15	10	5	4	14	11	2	7	15	10	3	6
7	2	16	9	7	2	16	9	6	3	16	9	6	3	16	9	4	5	16	9	4	5	16	9
14	11	5	4	12	13	3	6	15	10	5	4	12	13	2	7	15	10	3	6	14	11	2	7

1	12	6	15	1	12	7	4	1	12	13	8	1	12	13	8	1	14	7	12	1	14	11	8
14	7	9	4	15	6	9	4	14	7	2	11	15	6	3	10	15	4	9	6	15	4	5	10
11	2	16	5	10	3	16	5	4	9	16	5	4	9	16	5	10	5	16	3	6	9	16	3
8	13	3	10	8	13	2	11	15	6	3	10	14	7	2	11	8	11	2	13	12	7	2	13

2	7	9	16	2	7	9	16	2	7	12	13	2	7	12	13	2	7	14	11	2	7	14	11
11	14	4	5	13	12	6	3	11	14	1	8	16	9	6	3	13	12	1	8	16	9	4	5
8	1	15	10	8	1	15	10	5	4	15	10	5	4	15	10	3	6	15	10	3	6	15	10
13	12	6	3	11	14	4	5	16	9	6	3	11	14	1	8	16	9	4	5	13	12	1	8

2	11	5	16	2	11	8	13	2	11	14	7	2	11	14	7	2	13	8	11	2	13	12	7
14	7	9	4	16	5	10	3	13	8	1	12	16	5	4	9	16	3	10	5	16	3	6	9
15	6	12	1	9	4	15	6	3	10	15	6	3	10	15	6	9	6	15	4	5	10	15	4
3	10	8	13	7	14	1	12	16	5	4	9	13	8	1	12	7	12	1	14	11	8	1	14

3	6	9	16	3	6	12	13	3	6	15	10	3	6	15	10	3	10	15	6	3	10	8	13
13	12	7	2	16	9	7	2	13	12	1	8	16	9	4	5	13	8	11	2	16	5	11	2
8	1	14	11	5	4	14	11	2	7	14	11	2	7	14	11	12	1	14	7	9	4	14	7
10	15	4	5	10	15	1	8	16	9	4	5	13	12	1	8	6	15	4	9	6	15	1	12

3	10	15	6	3	10	15	6	3	13	8	10	3	13	12	6	4	5	10	15	4	5	11	14
13	8	1	12	16	5	4	9	16	2	11	5	16	2	7	9	14	11	8	1	15	10	8	1
2	11	14	7	2	11	14	7	9	7	14	4	5	11	14	4	7	2	13	12	6	3	13	12
16	5	4	9	13	8	1	12	6	12	1	15	10	8	1	15	9	16	3	6	9	16	2	7

4	5	16	9	4	5	16	9	4	9	6	15	4	9	7	14	4	9	16	5	4	9	16	5
14	11	2	7	15	10	3	6	14	7	12	1	15	6	12	1	14	7	2	11	15	6	3	10
1	8	13	12	1	8	13	12	11	2	13	8	10	3	13	8	1	12	13	8	1	12	13	8
15	10	3	6	14	11	2	7	5	16	3	10	5	16	2	11	15	6	3	10	14	7	2	11

4	14	7	9	4	14	11	5	5	4	14	11	5	4	15	10	6	3	13	12	6	3	16	9
15	1	12	6	15	1	8	10	16	9	7	2	16	9	6	3	15	10	8	1	15	10	5	4
10	8	13	3	6	12	13	3	3	6	12	13	2	7	12	13	4	5	11	14	1	8	11	14
5	11	2	16	9	7	2	16	10	15	1	8	11	14	1	8	9	16	2	7	12	13	2	7

5 x 5 magic squares

With 25 cells and the numbers 1 to 25 once each, the sums are each 65. There are 3600 possible, each of which can be rotated and reflected to give 8 variations. In spite of there being so many, they are not at all easy to find except by a set of systematic rules.

11	4	17	10	23
24	12	5	18	6
7	25	13	1	19
20	8	21	14	2
3	16	9	22	15

16	5	23	12	9
13	7	19	1	25
4	21	15	8	17
10	18	2	24	11
22	14	6	20	3

1	14	22	8	20
7	18	5	11	24
15	21	9	17	3
19	2	13	25	6
23	10	16	4	12

9	18	2	11	25
12	21	10	19	3
20	4	13	22	6
23	7	16	5	14
1	15	24	8	17

Concentric or embedded magic squares

It is possible to construct 5 x 5 magic squares so that the centre 3 x 3 square is also magic.

19	4	21	18	3
20	16	9	14	6
2	11	13	15	24
1	12	17	10	25
23	22	5	8	7

Totals of 3 x 3 square = 39
Totals of 5 x 5 square = 65

30	8	9	27	26
29	23	16	21	11
12	18	20	22	28
15	19	24	17	25
14	32	31	13	10

Totals of 3 x 3 square = 60
Totals of 5 x 5 square =100

1000 random numbers

9769977726	1085439714	3087410079	0553768187	8872686110
6393400611	5464877308	9820936418	2361578484	2322724992
1534998744	8700794242	9699352977	0766075501	3870382981
2824633096	1171258962	1886254239	2859767927	0868455661
4742587636	1359531625	3936638663	5905210047	8775926744
1698244704	7394022160	6383658596	7969470521	0376918022
9563535758	2632281344	2900531028	3534058467	4455361689
1691977276	5998373779	7141315207	7260404430	2517025309
4330399376	2175955706	1335103601	9089424841	2192762013
9974382145	6476385138	8213886932	4212376111	4228112016
0093440658	7613440824	6313957842	1856220099	8311509601
7335318978	0184623830	4407486322	9820716881	4430300792
9427099457	9283109250	8817176451	1751000285	4729569928
2907612191	1272061525	0702820975	4902873998	1445641359
5038614458	1084855818	3885505242	9764223026	5487511739
8300791411	2814556847	0672601585	0474634885	6274758657
5358719994	3382760579	6745258458	4065787449	8455999071
3098141584	3102088427	7981814043	3114754102	8195333553
1203400938	6204637897	5490182316	9793102157	6692156162
9638427820	3198331412	9190726761	7373768410	7157501405

NUMBER OF DIGITS

0	1	2	3	4	5	6	7	8	9
99	110	100	98	101	95	89	106	104	98

TOTAL 1000

Random numbers on a computer

Strictly speaking a computer cannot produce random numbers at all but can only do what it is programmed to do. Yet the instruction RND produces a string of apparently random numbers. How does it do it ?

Essentially the computer takes an input number and then performs some complicated calculation using it to produce a series of digits. Some computers always use the same input numbers and so the string of ´random´ numbers are always the same. Others count until a key is pressed and then use the number so reached as the input. It is interesting to test the RND function on different computers and then to get it to count the frequency of each digit and see how ´random´ they are. It is an easy matter to perform a chi-squared test on your results.

To do this,we must calculate

$$X^2 = \Sigma(O-E)^2/E$$

where O is the observed frequency and E the expected frequency.

For 1000 digits which are thought to be random,the expected value of E for each digit is 100. The value $(O-E)^2$ is the square of the difference between the observed value and 100.

For the random numbers listed above

$$X^2=(1^2+10^2+0^2+2^2+1^2+5^2+11^2+6^2+4^2+2^2)/100= 3.08$$

For 10 cells and 9 degrees of freedom, the chi-squared value is 16.92. Only if the calculated value of X^2 had been greater than 16.92 would we have doubted that the numbers are random. Of course, other tests are also needed, but they pass this first test.It is interesting to try the same test on π , e, $\sqrt{2}$, and the golden ratio on the 1000 digits of each given on pages 34 and 35. The values of X^2 are 4.74, 4.86, 8.38, 10.04 respectively and so all can be accepted as ´random´, in the digit sense, although each is a uniquely determined decimal, a property which can be used to produce random numbers on a computer.

π to 1000 decimal places

3.1415926535 8979323846 2643383279 5028841971 6939937510
5820974944 5923078164 0628620899 8628034825 3421170679
8214808651 3282306647 0938446095 5058223172 5359408128
4811174502 8410270193 8521105559 6446229489 5493038196
4428810975 6659334461 2847564823 3786783165 2712019091
4564856692 3460348610 4543266482 1339360726 0249141273
7245870066 0631558817 4881520920 9628292540 9171536436
7892590360 0113305305 4882046652 1384146951 9415116094
3305727036 5759591953 0921861173 8193261179 3105118548
0744623799 6274956735 1885752724 8912279381 8301194912
9833673362 4406566430 8602139494 6395224737 1907021798
6094370277 0539217176 2931767523 8467481846 7669405132
0005681271 4526356082 7785771342 7577896091 7363717872
1468440901 2249534301 4654958537 1050792279 6892589235
4201995611 2129021960 8640344181 5981362977 4771309960
5187072113 4999999837 2978049951 0597317328 1609631859
5024459455 3469083026 4252230825 3344685035 2619311881
7101000313 7838752886 5875332083 8142061717 7669147303
5982534904 2875546873 1159562863 8823537875 9375195778
1857780532 1712268066 1300192787 6611195909 2164201989

NUMBER OF DIGITS

0	1	2	3	4	5	6	7	8	9
93	116	103	102	93	97	94	95	101	106

TOTAL 1000

e to 1000 decimal places

2.7182818284 5904523536 0287471352 6624977572 4709369995
9574966967 6277240766 3035354759 4571382178 5251664274
2746639193 2003059921 8174135966 2904357290 0334295260
5956307381 3232862794 3490763233 8298807531 9525101901
1573834187 9307021540 8914993488 4167509244 7614606680
8226480016 8477411853 7423454424 3710753907 7744992069
5517027618 3860626133 1384583000 7520449338 2656029760
6737113200 7093287091 2744374704 7230696977 2093101416
9283681902 5515108657 4637721112 5238978442 5056953696
7707854499 6996794686 4454905987 9316368892 3009879312
7736178215 4249992295 7635148220 8269895193 6680331825
2886939849 6465105820 9392398294 8879332036 2509443117
3012381970 6841614039 7019837679 3206832823 7646480429
5311802328 7825098194 5581530175 6717361332 0698112509
9618188159 3041690351 5988885193 4580727386 6738589422
8792284998 9208680582 5749279610 4841984443 6346324496
8487560233 6248270419 7862320900 2160990235 3043699418
4914631409 3431738143 6405462531 5209618369 0888707016
7683964243 7814059271 4563549061 3031072085 1038375051
0115747704 1718986106 8739696552 1267154688 9570350354

NUMBER OF DIGITS

0	1	2	3	4	5	6	7	8	9
100	96	97	109	100	85	99	99	103	112

TOTAL 1000

√2 to 1000 decimal places

1.4142135623 7309504880 1688724209 6980785696 7187537694
8073176679 7379907324 7846210703 8850387534 3276415727
3501384623 0912297024 9248360558 5073721264 4121497099
9358314132 2266592750 5592755799 9505011527 8206057147
0109559971 6059702745 3459686201 4728517418 6408891986
0955232923 0484308714 3214508397 6260362799 5251407989
6872533965 4633180882 9640620615 2583523950 5474575028
7759961729 8355752203 3753185701 1354374603 4084988471
6038689997 0699004815 0305440277 9031645424 7823068492
9369186215 8057846311 1596668713 0130156185 6898723723
5288509264 8612494977 1542183342 0428568606 0146824720
7714358548 7415565706 9677653720 2264854470 1585880162
0758474922 6572260020 8558446652 1458398893 9443709265
9180031138 8246468157 0826301005 9485870400 3186480342
1948972782 9064104507 2636881313 7398552561 1732204024
5091227700 2269411275 7362728049 5738108967 5040183698
6836845072 5799364729 0607629969 4138047565 4823728997
1803268024 7442062926 9124859052 1810044598 4215059112
0249441341 7285314781 0580360337 1077309182 8693147101
7111168391 6581726889 4197587165 8215212822 9518488472

NUMBER OF DIGITS

0	1	2	3	4	5	6	7	8	9
108	98	109	82	100	104	90	104	113	92

TOTAL 1000

The golden ratio to 1000 decimal places

1.6180339887 4989484820 4586834365 6381177203 0917980576
2862135448 6227052604 6281890244 9707207204 1893911374
8475408807 5386891752 1266338622 2353693179 3180060766
7263544333 8908659593 9582905638 3226613199 2829026788
0675208766 8925017116 9620703222 1043216269 5486262963
1361443814 9758701220 3408058879 5445474924 6185695364
8644492410 4432077134 4947049565 8467885098 7433944221
2544877066 4780915884 6074998871 2400765217 0575179788
3416625624 9407589069 7040002812 1042762177 1117778053
1531714101 1704666599 1466979873 1761356006 7087480710
1317952368 9427521948 4353056783 0022878569 9782977834
7845878228 9110976250 0302696156 1700250464 3382437764
8610283831 2683303724 2926752631 1653392473 1671112115
8818638513 3162038400 5222165791 2866752946 5490681131
7159934323 5973494985 0904094762 1322298101 7261070596
1164562990 9816290555 2085247903 5240602017 2799747175
3427775927 7862561943 2082750513 1218156285 5122248093
9471234145 1702237358 0577278616 0086883829 5230459264
7878017889 9219902707 7690389532 1968198615 1437803149
9741106926 0886742962 2675756052 3172777520 3536139362

NUMBER OF DIGITS

0	1	2	3	4	5	6	7	8	9
100	105	116	88	92	84	104	113	105	93

TOTAL 1000

Numbers in base 10

Ordinary numbers make use of ten different signs to represent the numbers 0,1,2,3,4,5,6,7,8,9. The next number is then 10, followed by 11,12 etc.

LIST OF NUMBERS IN BASE 10 UP TO 100

1	2	3	4	5	6	7	8	9	10
11	12	13	14	15	16	17	18	19	20
21	22	23	24	25	26	27	28	29	30
31	32	33	34	35	36	37	38	39	40
41	42	43	44	45	46	47	48	49	50
51	52	53	54	55	56	57	58	59	60
61	62	63	64	65	66	67	68	69	70
71	72	73	74	75	76	77	78	79	80
81	82	83	84	85	86	87	88	89	90
91	92	93	94	95	96	97	98	99	100

MULTIPLICATION TABLES IN BASE 10.

Although in Britain it is the convention to learn multiplication tables as far as 12x12, it is not necessary to do so in order to be able to multiply and divide.

	1	2	3	4	5	6	7	8	9
1	1	2	3	4	5	6	7	8	9
2	2	4	6	8	10	12	14	16	18
3	3	6	9	12	15	18	21	24	27
4	4	8	12	16	20	24	28	32	36
5	5	10	15	20	25	30	35	40	45
6	6	12	18	24	30	36	42	48	54
7	7	14	21	28	35	42	49	56	63
8	8	16	24	32	40	48	56	64	72
9	9	18	27	36	45	54	63	72	81

Numbers in bases 9 and 8

BASE 9. There are 9 digits which we may call 0,1,2,3,4,5,6,7,8

LIST OF NUMBERS TO 100 IN BASE 9

1	2	3	4	5	6	7	8	10
11	12	13	14	15	16	17	18	20
21	22	23	24	25	26	27	28	30
31	32	33	34	35	36	37	38	40
41	42	43	44	45	46	47	48	50
51	52	53	54	55	56	57	58	60
61	62	63	64	65	66	67	68	70
71	72	73	74	75	76	77	78	80
81	82	83	84	85	86	87	88	100

MULTIPLICATION TABLES IN BASE 9

	1	2	3	4	5	6	7	8
1	1	2	3	4	5	6	7	8
2	2	4	6	8	11	13	15	17
3	3	6	10	13	16	20	23	26
4	4	8	13	17	22	26	31	35
5	5	11	16	22	27	33	38	44
6	6	13	20	26	33	40	46	53
7	7	15	23	31	38	46	54	62
8	8	17	26	35	44	53	62	71

BASE 8. There are 8 digits which we may call 0,1,2,3,4,5,6,7

LIST OF NUMBERS TO 100 IN BASE 8

1	2	3	4	5	6	7	10
11	12	13	14	15	16	17	20
21	22	23	24	25	26	27	30
31	32	33	34	35	36	37	40
41	42	43	44	45	46	47	50
51	52	53	54	55	56	57	60
61	62	63	64	65	66	67	70
71	72	73	74	75	76	77	100

MULTIPLICATION TABLES IN BASE 8

	1	2	3	4	5	6	7
1	1	2	3	4	5	6	7
2	2	4	6	10	12	14	16
3	3	6	11	14	17	22	25
4	4	10	14	20	24	30	34
5	5	12	17	24	31	36	43
6	6	14	22	30	36	44	52
7	7	16	25	34	43	52	61

Numbers in bases 7,6,5,4,3,2

BASE 7. There are 7 digits which may be called 0,1,2,3,4,5,6

LIST OF NUMBERS to 100

1	2	3	4	5	6	10
11	12	13	14	15	16	20
21	22	23	24	25	26	30
31	32	33	34	35	36	40
41	42	43	44	45	46	50
51	52	53	54	55	56	60
61	62	63	64	65	66	100

MULTIPLICATION TABLES

	1	2	3	4	5	6
1	1	2	3	4	5	6
2	2	4	6	11	13	15
3	3	6	12	15	21	24
4	4	11	15	22	26	33
5	5	13	21	26	34	42
6	6	15	24	33	42	51

BASE 6. There are 6 digits which may be called 0,1,2,3,4,5

LIST OF NUMBERS TO 100

1	2	3	4	5	10
11	12	13	14	15	21
21	22	23	24	25	30
31	32	33	34	35	40
41	42	43	44	45	50
51	52	53	54	55	100

MULTIPLICATION TABLES

	1	2	3	4	5
1	1	2	3	4	5
2	2	4	10	12	14
3	3	10	13	20	23
4	4	12	20	24	32
5	5	14	23	32	41

BASE 5. There are 5 digits which may be called 0,1,2,3,4

LIST OF NUMBERS TO 100

1	2	3	4	10
11	12	13	14	20
21	22	23	24	30
31	32	33	34	40
41	42	43	44	100

MULTIPLICATION TABLES

	1	2	3	4
1	1	2	3	4
2	2	4	11	13
3	3	11	14	22
4	4	13	22	31

BASE 4. There are 4 digits which may be called 0,1,2,3,

LIST OF NUMBERS TO 100

1	2	3	10
11	12	13	20
21	22	23	30
31	32	33	100

MULTIPLICATION TABLES

	1	2	3
1	1	2	3
2	2	10	12
3	3	12	21

BASE 3. There are 3 digits which may be called 0,1,2,

LIST OF NUMBERS TO 100

1	2	10
11	12	20
21	22	100

MULTIPLICATION TABLES

	1	2
1	1	2
2	2	11

BASE 2. There are 2 digits which may be called 0,1,

LIST OF NUMBERS TO 100

1	10
11	100

MULTIPLICATION TABLES

	1
1	1

Numbers in base 12

Once the number of digits required is more than 10, we need to invent new symbols to represent the extra ones. It is often convenient to use letters of the alphabet.

BASE 12. There are 12 digits which we may call 0,1,2,3,4,5,6,7,8,9,A,B,

This system of numbers is known as DUODECIMAL and offers some advantage over base 10. For instance, in duodecimal 3x4=10 and so the base has factors of 1,2,3,4,(but not 5). It is a base which might well have become the standard one, but through possibly a historical accident did not.In many countries the idea of counting in ´dozens´is widespread. Things pack conveniently in dozens and so it is rare to buy eggs or bottles of wine in tens. The British Imperial system of weights and measures make more use of twelve than ten and even today children learn their muliplication tables as far as 12 x 12 when 9 x 9 would suffice.

LIST OF NUMBERS TO 100 IN BASE 12

1	2	3	4	5	6	7	8	9	A	B	10
11	12	13	14	15	16	17	18	19	1A	1B	20
21	22	23	24	25	26	27	28	29	2A	2B	30
31	32	33	34	35	36	37	38	39	3A	3B	40
41	42	43	44	45	46	47	48	49	4A	4B	50
51	52	53	54	55	56	57	58	59	5A	5B	60
61	62	63	64	65	66	67	68	69	6A	6B	70
71	72	73	74	75	76	77	78	79	7A	7B	80
81	82	83	84	85	86	87	88	89	8A	8B	90
91	92	93	94	95	96	97	98	99	9A	9B	A0
A1	A2	A3	A4	A5	A6	A7	A8	A9	AA	AB	B0
B1	B2	B3	B4	B5	B6	B7	B8	B9	BA	BB	100

MULTIPLICATION TABLES IN BASE 12

	1	2	3	4	5	6	7	8	9	A	B
1	1	2	3	4	5	6	7	8	9	A	B
2	2	4	6	8	A	10	12	14	16	18	1A
3	3	6	9	10	13	16	19	20	23	26	29
4	4	8	10	14	18	20	24	28	30	34	38
5	5	A	13	18	21	26	2B	34	39	42	47
6	6	10	16	20	26	30	36	40	46	50	56
7	7	12	19	24	2B	36	41	48	53	5A	65
8	8	14	20	28	34	40	48	54	60	68	74
9	9	16	23	30	39	46	53	60	69	76	83
A	A	18	26	34	42	50	5A	68	76	84	92
B	B	1A	29	38	47	56	65	74	83	92	A1